Jesse Herman Holmes

Chichen Itza, Mexico, 1895

Jesse Herman Holmes

Chichen Itza, Mexico, 1895

ISBN/EAN: 9783743432758

Manufactured in Europe, USA, Canada, Australia, Japa

Cover: Foto ©berggeist007 / pixelio.de

Manufactured and distributed by brebook publishing software
(www.brebook.com)

Jesse Herman Holmes

Chichen Itza, Mexico, 1895

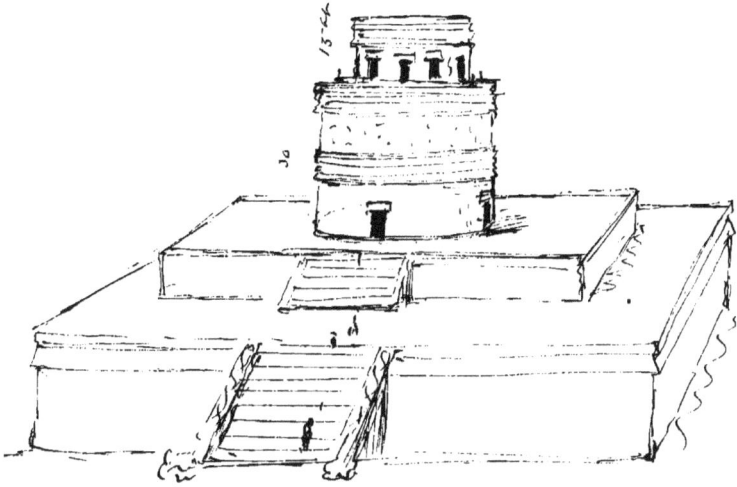

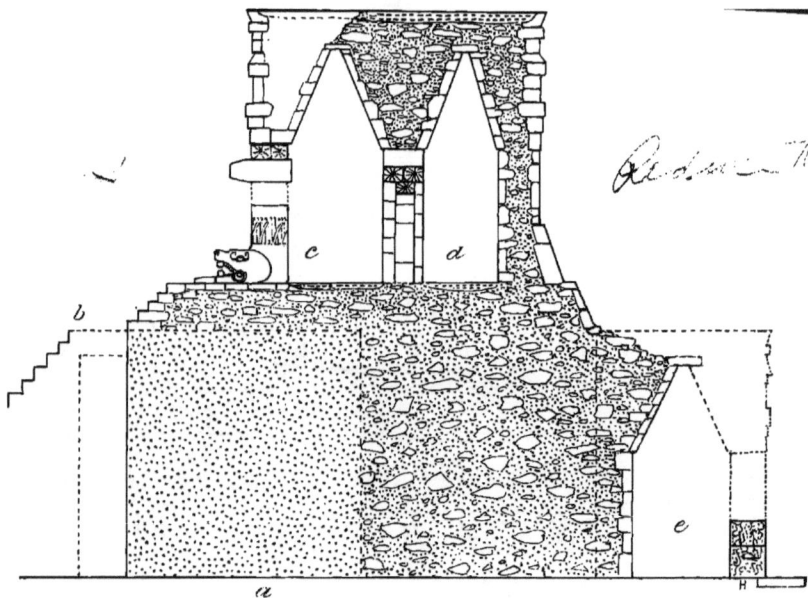

1970-31-10261

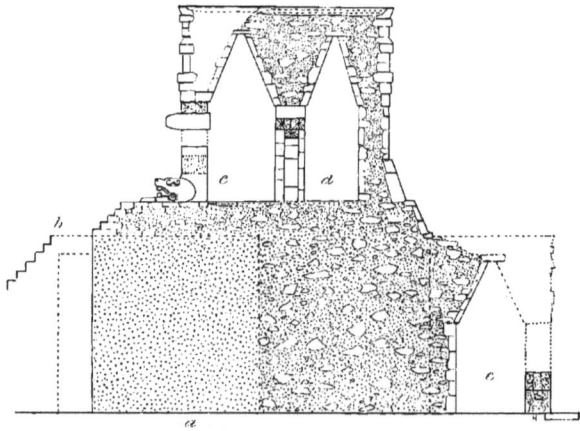

Fig. 51. Section of temple of Sigus showing unique entablature profile.
Parts not filled in with are older restorations.

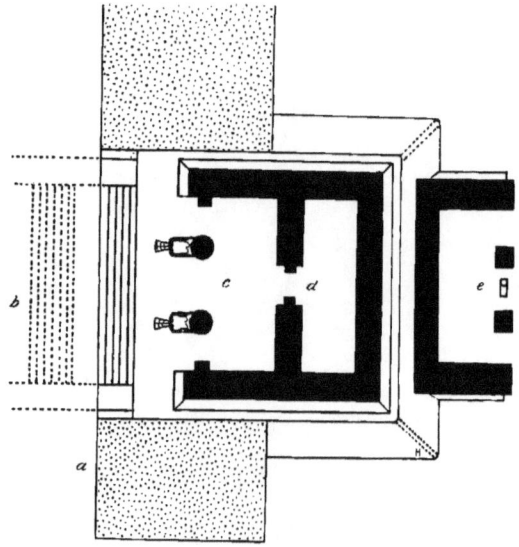

Fig. 52. Plan of temple of Sigus, restored on the lines now existing. b

Fig. 70. Section of Nunnery or Palace Chichen Itza, showing earliest portion *a*, with superstructures and additions at the right side and left.

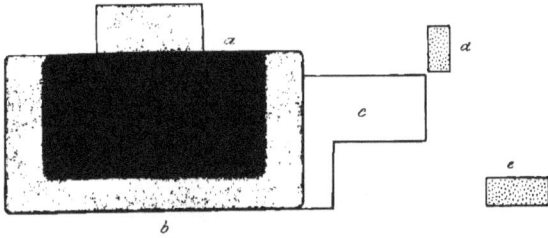

Fig 37. Plan of monastery palace, indicating the side stage of growth. The black represents the nucleus, the gray the first addition, the white the additions.

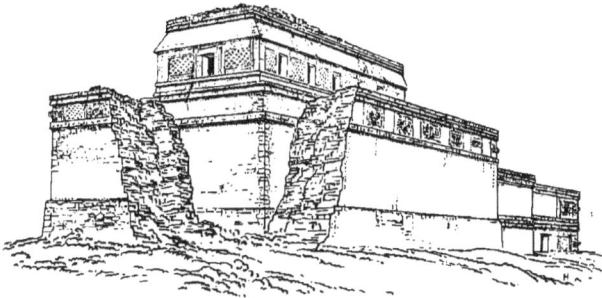

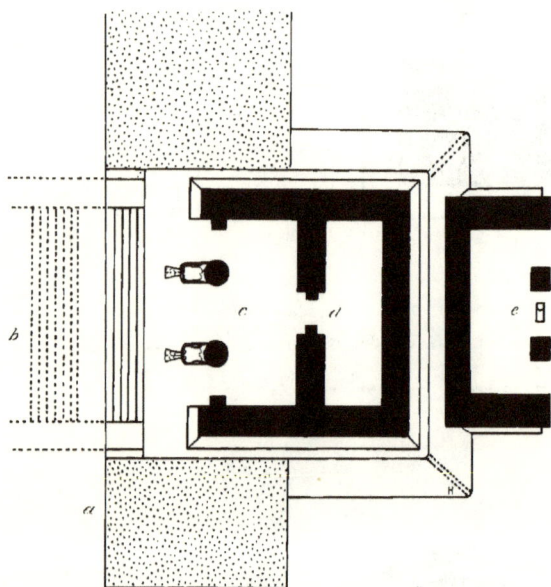

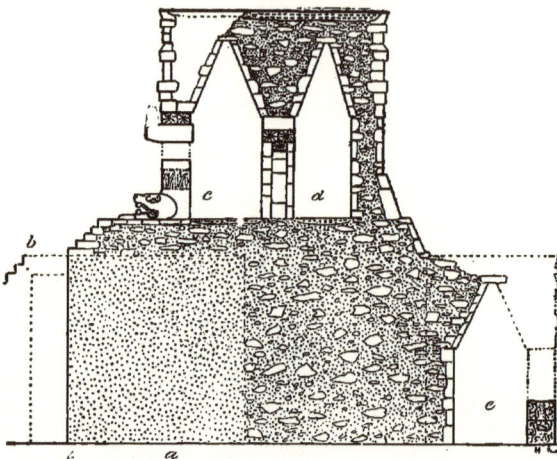

FIG. 38. SECTION OF TEMPLE OF TIGERS, SHOWING UNIQUE ENTABLATURE PROFILE.

Façade of upper structure and front face of vestibule vault above the middle of the column are
restored. The lines should have been uniformly dotted. Façade of lower structure
may not have occurred as here indicated, for there is a possibility that
other chambers existed to the right of this.

a. Section of Gymnasium wall—dotted area.
b. Hypothetic continuation of stairway.
c. Serpent column and outer chamber or vestibule.
d. Sanctuary, entered by pilastered doorway with three wooden lintels.
e. Lower Temple, with sculptured columns and walls.

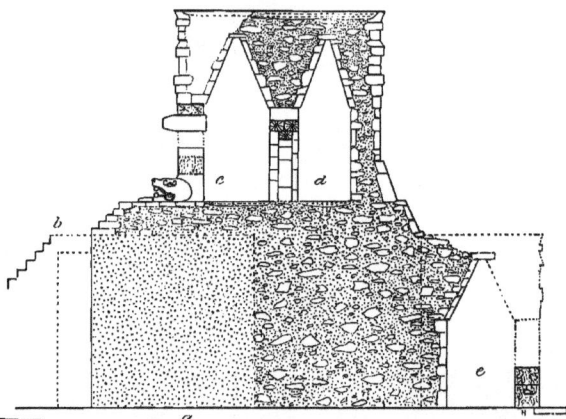

FIG. 1. SECTION OF TEMPLE OF TIGERS SHOWING UNIQUE ENTABLATURE
PROFILE.

Facades
~~Front~~ ~~Portrns~~ with Tumbles and the front face of the Corridor vaults above the middle within the cides

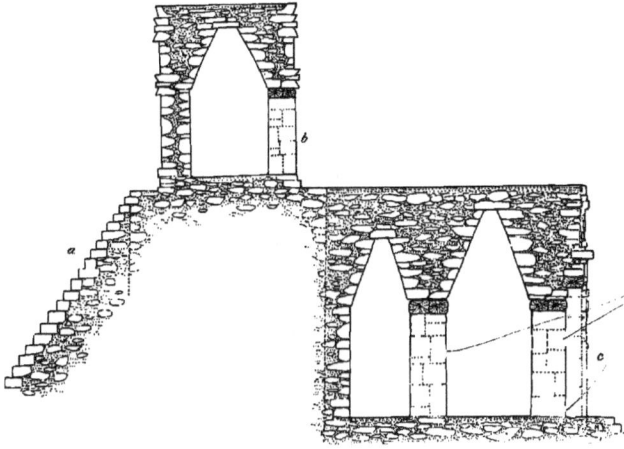

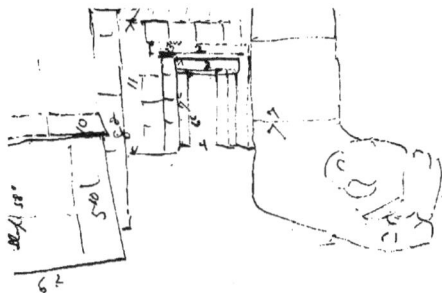

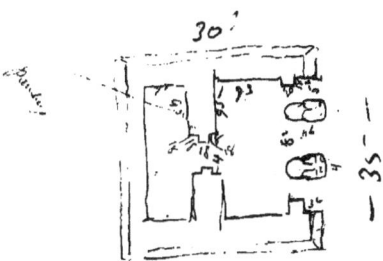

30'

— 35' —

Jacob
Kofel

Ground vault painted
all over about 24ft high

Miscellany

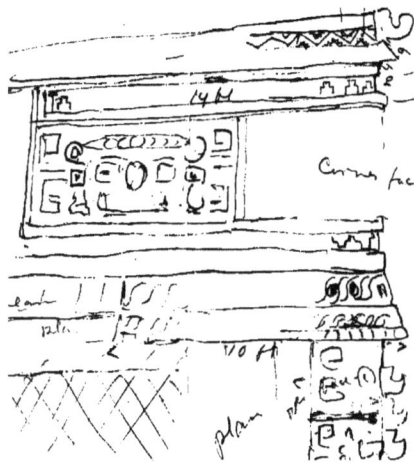

Cornice fac

east

plan

14 ft

10 ft

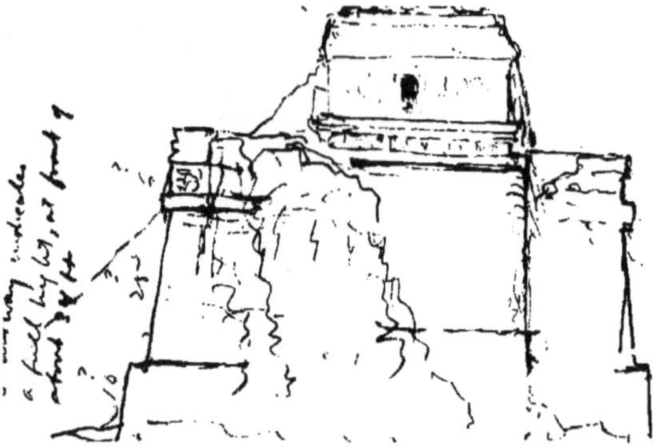

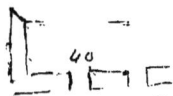

Back of the room

40

16

3

6

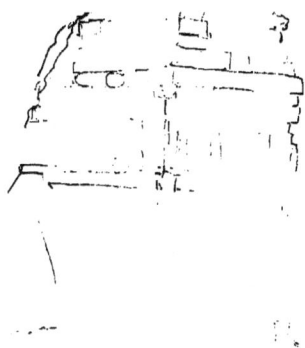

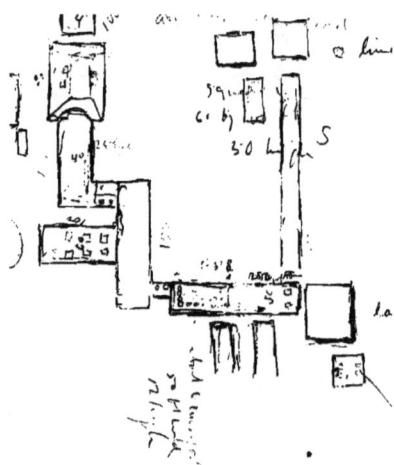

S

N

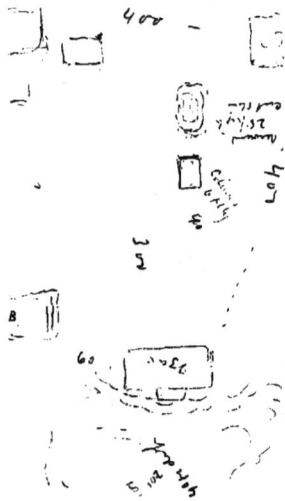

En

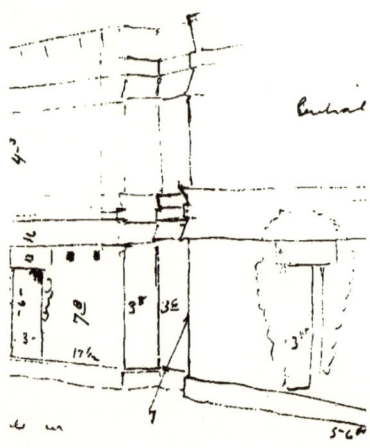

plastered wall

arch

wall

20

West

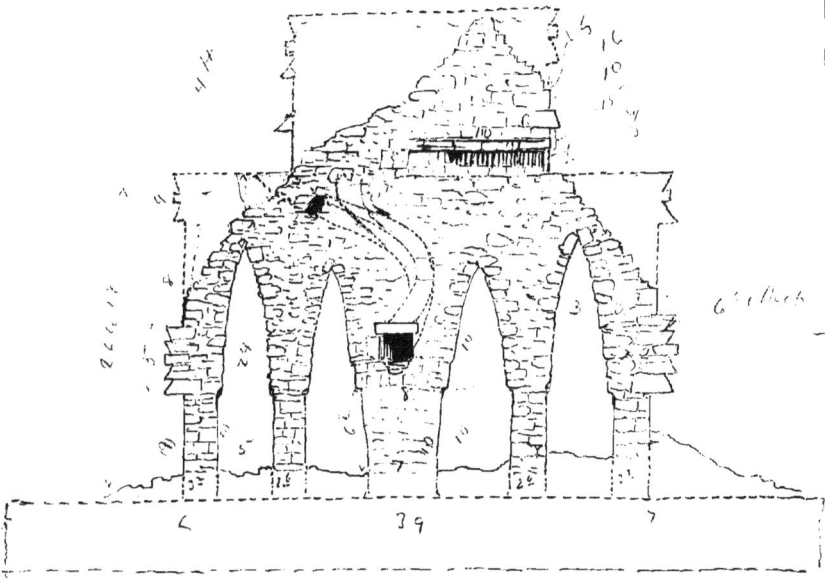

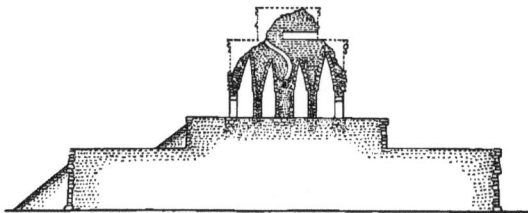

Fig. 43. Section of Round Tower or Caracol with its terrace and hypothetic profile of ~~summit~~ upper portion of turrets.

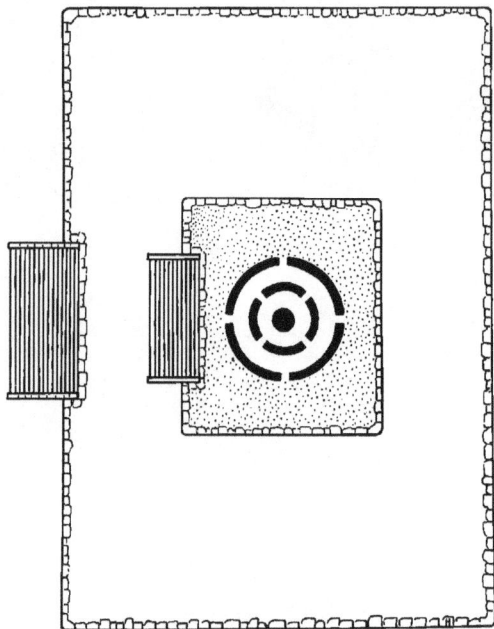

Profil of races Lorraine.

course of the masonry

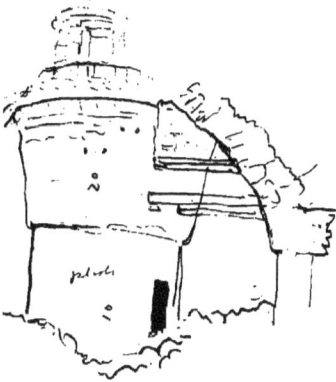

·/ outli doom 3'4 wide

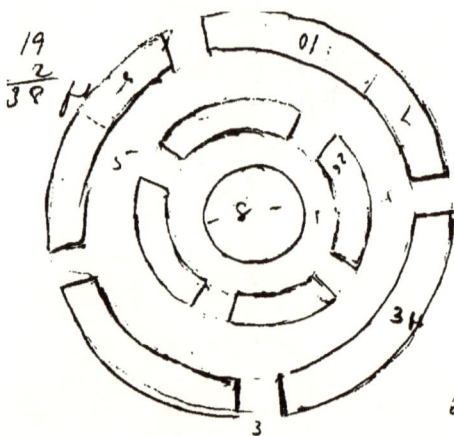

$\dfrac{19}{3\,8}\ \dfrac{2}{}$

01

3

3

openings in each
means equidistant
in two walls are
...

Diam. 38 to 40. ft.

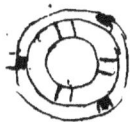

outer arch 24 ft
high
5 ft wide at base

The second along, the portion showing
5 doors in the front face, is the
visible portion of the earliest building
& is exposed to nearly the full height
at the N.W. Corner where the massive
terrace is broken away

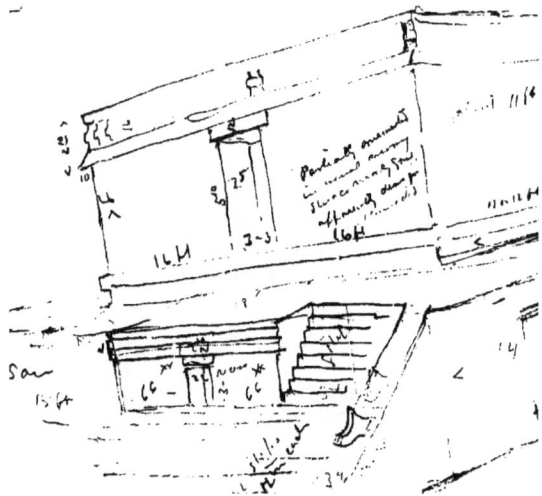

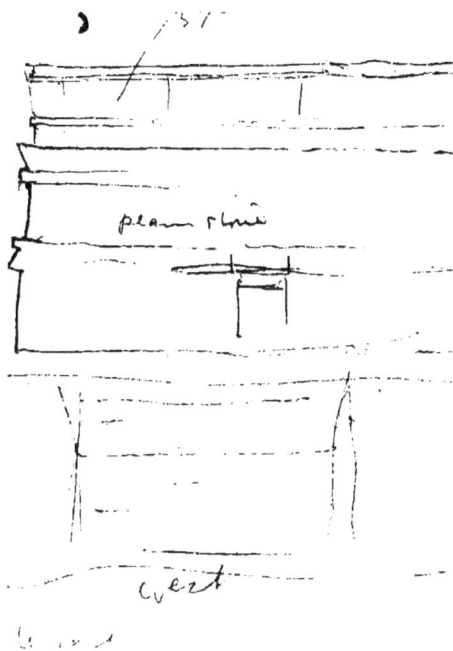

plan shui

cvert

Next Round to
west

"The prism"
Jan 17

all plan walls

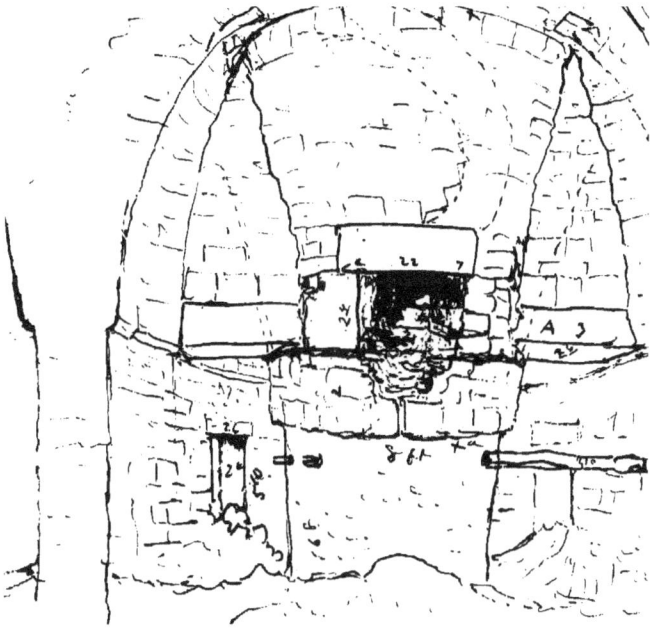

Round tower

\

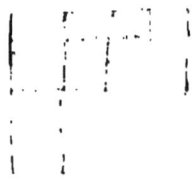

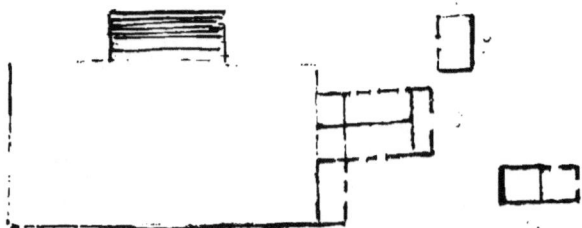

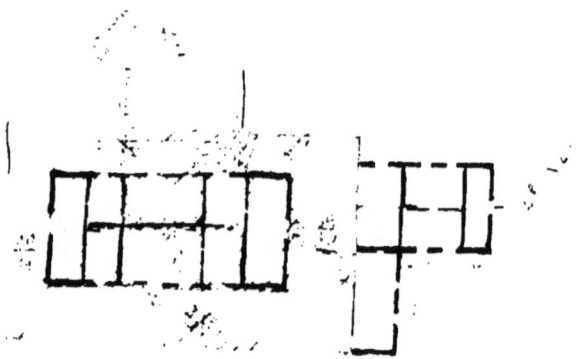